ULTIMATE SUPERCARS

ASTON MARTIN VALKYRIE

By James Savino

Kaleidoscope
Minneapolis, MN

Bigfoot Books

The Quest for Discovery Never Ends

....

This edition first published in 2023 by Kaleidoscope Publishing, Inc.

No part of this publication may be reproduced in whole or in part without written permission of the publisher.

For information regarding permission, write to
Kaleidoscope Publishing, Inc.
6012 Blue Circle Drive
Minnetonka, MN 55343

Library of Congress Control Number
2022937978

ISBN
978-1-64519-605-1 (library bound)
978-1-64519-675-4 (ebook)

Text copyright © 2023 by Kaleidoscope Publishing, Inc. All-Star Sports, Bigfoot Books, and associated logos are trademarks and/or registered trademarks of Kaleidoscope Publishing, Inc.

Printed in the United States of America.

Bigfoot lurks within one of the images in this book. It's up to you to find him!

TABLE OF
CONTENTS

Chapter 1: Road or Racetrack? **4**

Chapter 2: Named for a Hill **12**

Chapter 3: All About Speed **18**

Chapter 4: Hard to Find **24**

Beyond the Book *28*
Research Ninja *29*
Further Resources *30*
Glossary *31*
Index *32*
Photo Credits *32*
About the Author *32*

Chapter 1
Road or Racetrack?

Holley was blown away! She couldn't believe her eyes. The new 2022 Aston Martin Valkyrie supercars poster was on her bedroom wall. She lay on her bed, staring at the poster. The cars seemed to go fast even though they were sitting still. Just then, there was a knock at her door. "Come in."

It was Holley's brother, Norman. "What do you think of the poster, kid?"

"It's the best. One day I'll be driving one of those cars."

"You'll be a race car driver, and I'll be James Bond driving an Aston Martin DB5," said Norman.

"Deal!" said Holley.

The Valkyrie comes in three models. The **hardtop** Coupe, **convertible** Spider, and a track-only AMR PRO. Holley dreams of driving the Spider. She wants the world to see her driving a supercar.

At dinner, Holley asked, "Have you heard of Red Bull Racing?"

"It's one of the top Formula One racing teams," said her dad. "They partnered with Aston Martin to design the Valkyrie."

Norman sat down at the table. "The **aerodynamic** body style with the mid-engine design is cool."

MIDDLE MOTOR

Mid-engine cars have the engine behind the driver. Drivers like Valkyrie's mid-engine. It gives the cars better handling on the street and racetrack. SUVs and cars' engines are usually in front of the driver.

"What about the **V-12** engine? It has 1,000 **horsepower**! And the electric motor adds a 160-horsepower boost." Holley added.

The Valkyrie's teardrop-shaped **cockpit** doesn't have a back window. Instead, the driver has a rearview camera. The dashboard is sleek with a display for the camera. Front and rear parking sensors keep the driver from bumping the curb when parking.

Holley is curious about a feature outside the car that looks like huge tunnels. The **Venturi tunnels** are under the cockpit floor. Norman is taking engineering in school. He explains to Holley how they draw in big chunks of air. The air works with gravity to create **downforce**. This helps keep the Valkyries stable at high speeds.

PARTS OF A VALKYRIE

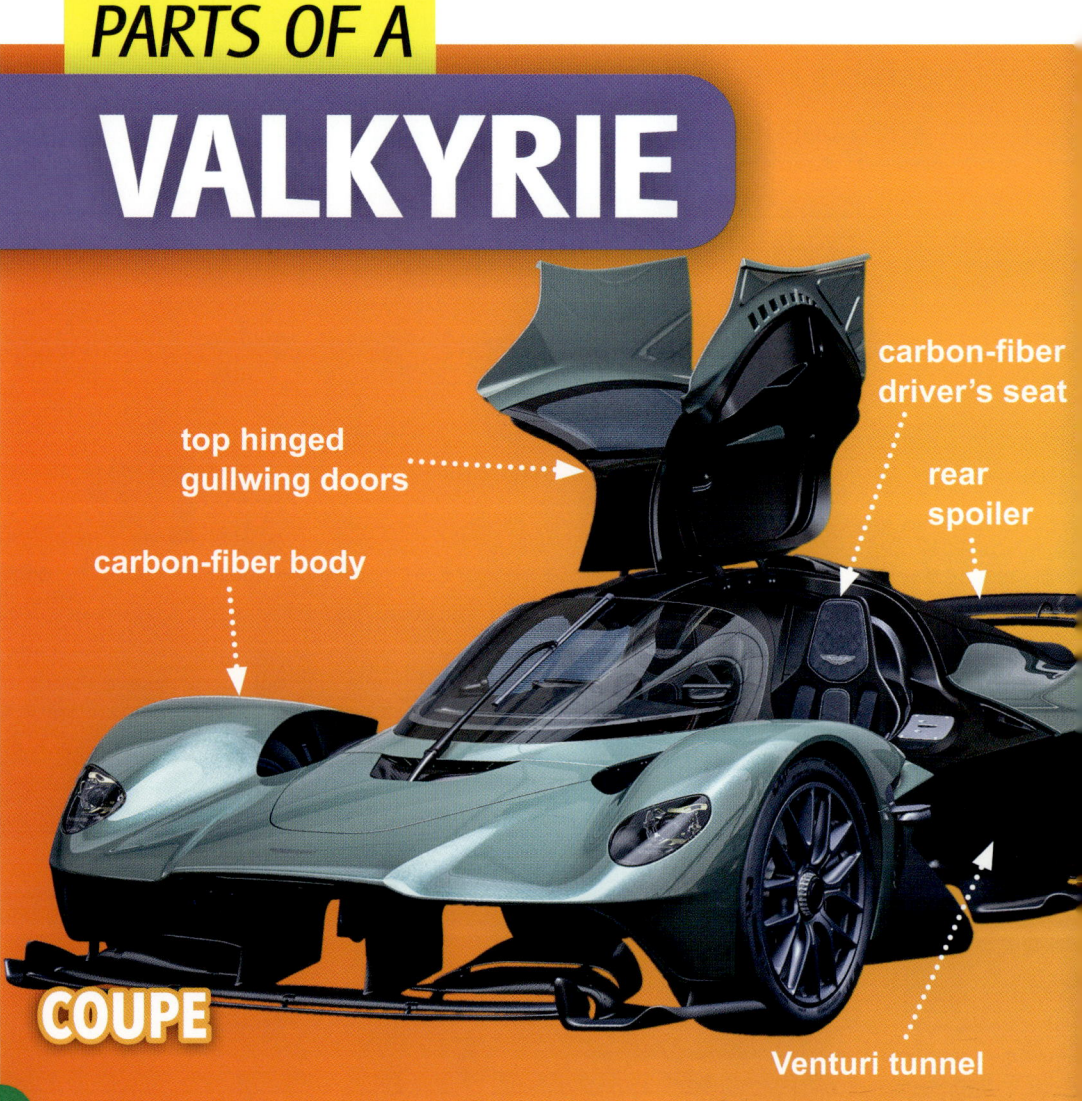

top hinged gullwing doors

carbon-fiber body

carbon-fiber driver's seat

rear spoiler

Venturi tunnel

COUPE

The Valkyrie Spider's **carbon-fiber** body is light. It doesn't have any steel. One way to tell the Spider and the Coupe apart is the doors. The Spider's doors are front-hinged. The Coupe's doors have roof hinges. Holley imagines herself behind the wheel, hearing the engine roar to life. She can't wait to get her driver's license.

front-hinged door

removable roof

cockpit

SPIDER

The first time the public experienced the Valkyrie was on January 10, 2022, at the British Grand Prix in England. Chris Godwin, Aston Martin's head test driver, drove one lap around the Silverstone Circuit. Aston Martin CEO Andy Palmer commented that the sound of the screaming V-12 and its accompanying hybrid system was one of the best sounds he'd ever heard.

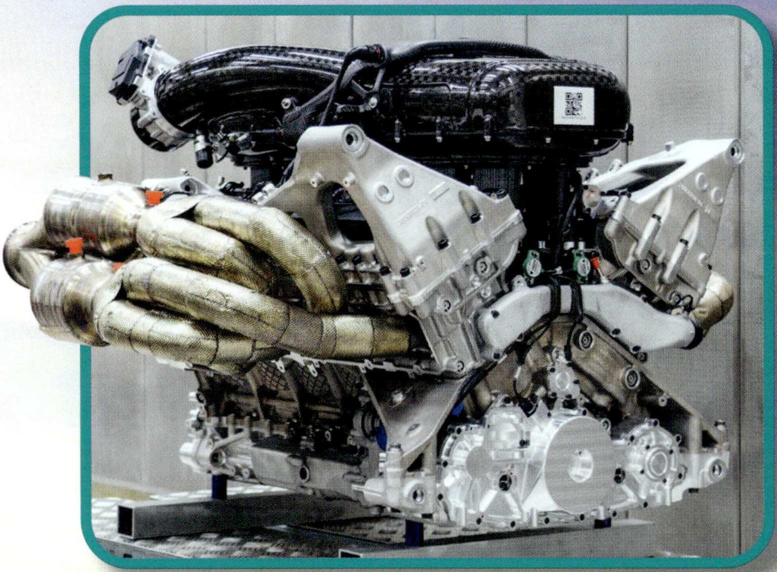

Chapter 2
Named for a Hill

Lionel Martin and Robert Bamford founded Aston Martin in 1913 in London. Bamford was an engineer. Martin was an engineer and a race car driver. In April 1914, Martin raced up a hill called Aston Hill. After that, the company was known as Aston Martin. Soon Aston Martin was racing cars at the highest levels of racing.

Aston Martin is one of the oldest carmakers in England. It has always been involved in racing.

ASTON MARTIN TIMELINE

1913 — Robert Bamford and Lionel Martin form Bamford and Martin LTD, based in Henniker Mews, off Fulham Road in London.

1915 — The first Aston Martin is registered on March 16. It's christened Coal Scuttle.

1922 — An Aston Martin nicknamed Bunny breaks ten world records in 16.5 hours at Brooklands.

1935 — The Aston Martin Owners Club is started at The Grafton Hotel in London.

1948 — The Sports wins the Spa-Francorchamps 24-hour race.

1958 — New model DB4 is launched.

1964 — James Bond drives the new DB5 in the movie *Goldfinger* and an iconic on-screen relationship is born.

1986 — The Vantage Zagato is launched and becomes one of the fastest supercars.

2005 — The DBR9 returns to the racetrack and secures victories at LeMans in 2007 and 2008.

2012 — The V-12 Vanquish returns as the flagship model.

Holley lives a few hours from the Aston Martin **headquarters**. But she's never been there. So maybe Norman would take her for her birthday. Norman agrees to take Holley. After all, he's an Aston Martin fan, too.

Fingers crossed, Holley will see a Valkyrie.

ASTON MARTIN HEADQUARTERS

Gaydon, Warwickshire, England

ULTRA-LUXURY

The Aston Martin headquarters and manufacturing plant are in the Gaydon Village of Warwickshire, England. Aston Martin's vision is to be the world's most desirable ultra-luxury British brand.

The tour guide said, "You two are very lucky to see a Valkyrie in production! Each car is made by hand. The robots only lift or move parts."

"How many Valkyries are you making?" asked Holley.

"We only make 150 road cars. So the Valkyrie is very desirable. Even with a price tag of over 3 million dollars. Most people who buy Valkyries keep them as collector's items. They rarely drive them."

Holley said, "I would drive mine every day!"

Norman laughed. "So would I, Holley, so would I."

"I hope you enjoyed your tour and have a happy birthday," said the tour guide.

"Thank you," said Norman and Holley.

Holley turned to Norman. "You're the best big brother. And this trip was the best birthday present ever!"

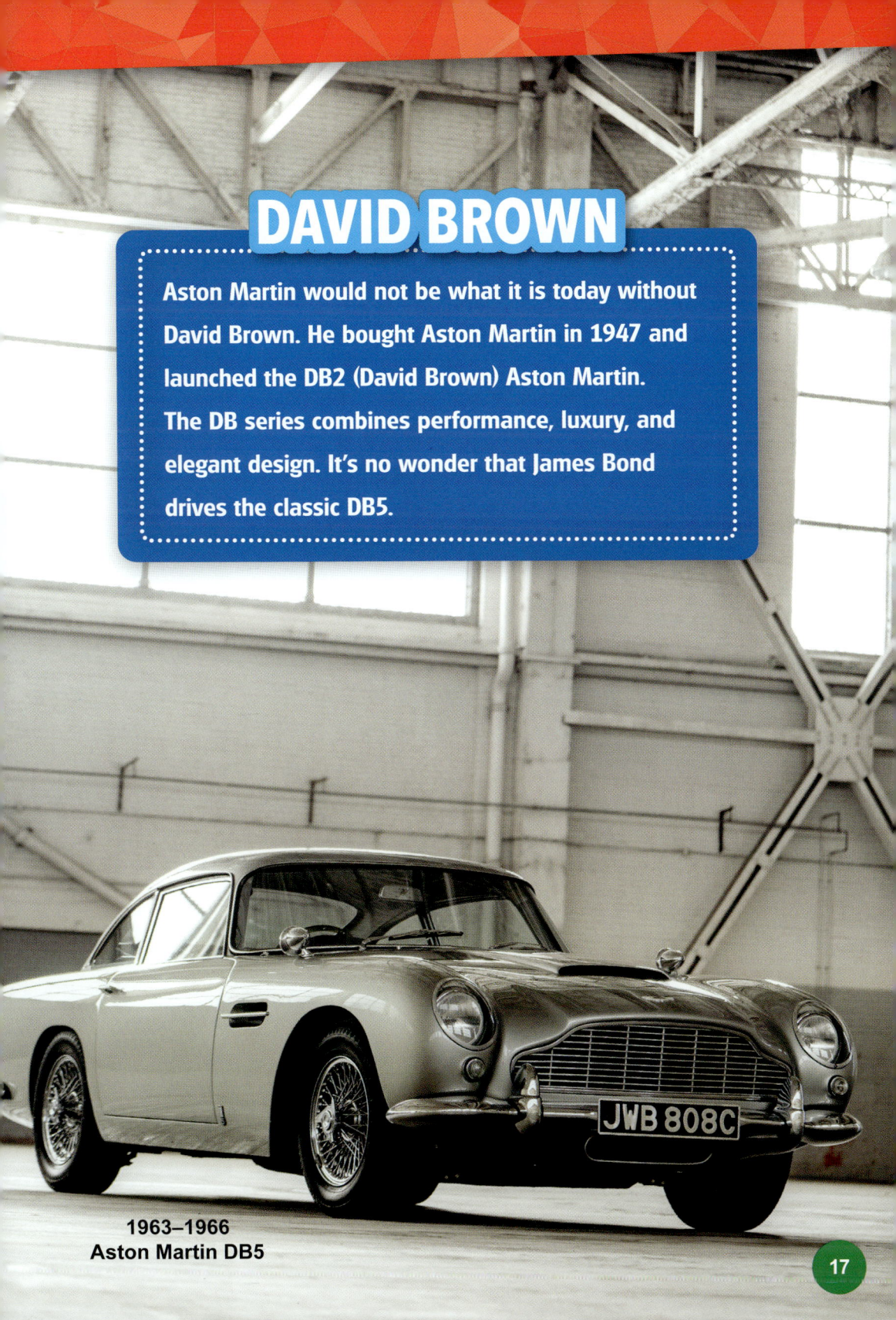

DAVID BROWN

Aston Martin would not be what it is today without David Brown. He bought Aston Martin in 1947 and launched the DB2 (David Brown) Aston Martin. The DB series combines performance, luxury, and elegant design. It's no wonder that James Bond drives the classic DB5.

1963–1966
Aston Martin DB5

Chapter 3
All About Speed

The Valkyrie is made with one purpose in mind—speed. There are no cup holders or even a radio. It's a race car for the streets. The Valkyrie can go from 0 to 200 miles per hour (322 km/h) in 10 seconds. It can also go from 200 to 0 miles per hour (322 km/h) in 5 seconds. The Formula One brake system makes it stop quickly.

STRONG LEGS HELP!

Many people don't think of race car drivers as athletes, but they are. Besides F-1 braking systems and downforce, strong legs are important in braking. Drivers must push the brake pedal, or the brakes won't work!

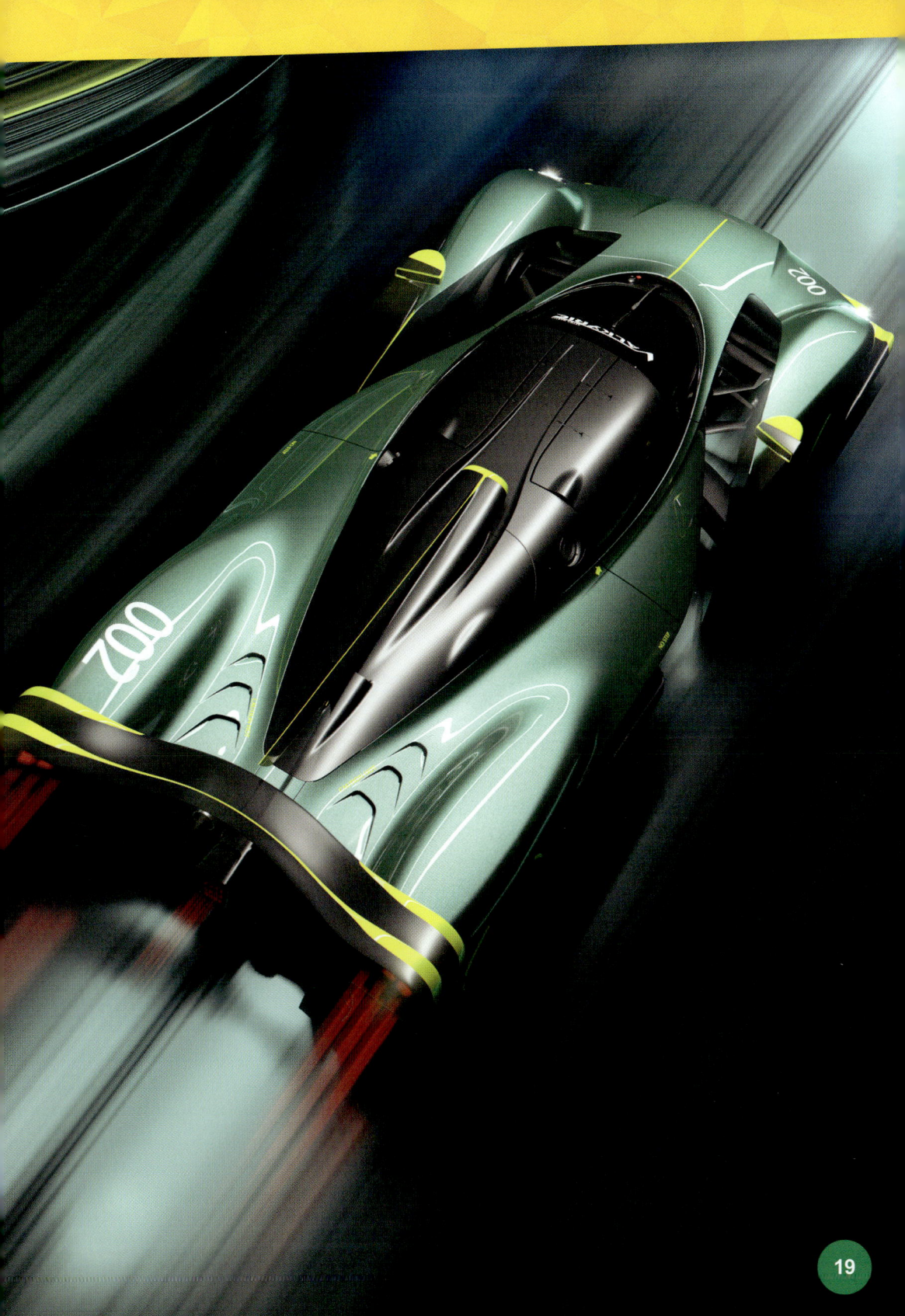

Unlike other hybrids, Valkyrie's V-12 engine uses the electric motor for a power boost. The transmission transfers power to the rear wheels. Holley thought it was nothing at all like her mom's Prius hybrid.

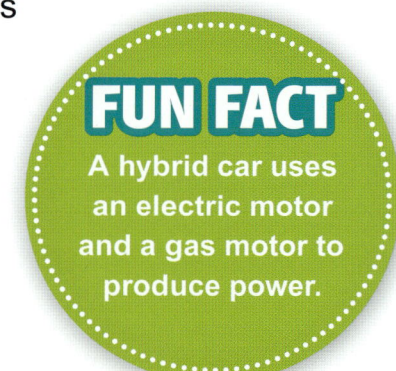

FUN FACT
A hybrid car uses an electric motor and a gas motor to produce power.

LET'S COMPARE

	PRIUS	VALKYRIE
0-60 miles per hour (95 km/h)	10.5 seconds	2.5 seconds
HORSEPOWER	121	1160
PRICE	$25,650	$3,000,000+

THE COUPE IN DETAIL

COST: $3,000,000 basic model

Height: 3.3 feet (1.0 m)

Width: 6.4 feet (1.95 m)

LENGTH: 15.8 feet (4.8 m)

WEIGHT: 3,300 pounds (1,497 kg)

TOP SPEED: 250 miles per hour (402 km/h)

TIME FROM 0 to 60 miles per hour: 2.5 seconds

FUN FACT

Aston Martin takes the safety of their drivers seriously. Instead of a seatbelt, the Valkyrie has a 6-point safety harness.

Chapter 4
Hard to Find

Aston Martin's racing history made the Valkyrie a collectible supercar before it came out of production. Supercar fans worldwide are excited for the 2022 Coupe, Spider, and AMR PRO.

Most Valkyries will not see the streets. Instead, their owners will drive them on racetracks. They want to experience the true supercar's exciting features. The 25 AMR PRO models will set records and win many Formula One races.

Holley was excited! She passed her driver's license test. Sitting behind the wheel of her mom's Prius hybrid, she imagined herself driving a Valkyrie. She thought about how it would handle sharp curves and winding roads. One day, she'll drive her dream car, the Valkyrie Spider!

BEYOND THE BOOK

After reading the book, it's time to think about what you learned. Try the following exercises to jump-start your ideas.

THINK

FIND OUT MORE. Think about what types of sources you could find on the Aston Martin Valkyrie. What could you find in a magazine? What could you learn at a dealership? How could each of the sources be useful in its own way?

CREATE

GET ARTISTIC. Use a primary resource, such as a photo, interview with an owner, or original document, to learn more about the Valkyrie. What new information do you want to learn about the Valkyrie? How can you locate the information?

SHARE

DIG DEEPER. Create a sales brochure for the Valkyrie. What important information would you include in the brochure? What images would you use that would interest your buyer?

GROW

GO TO A CAR SHOW. With an adult, see if you can visit a car dealership that carries the Valkyrie or other supercars. What would you hope to see or learn about the car?

RESEARCH NINJA

Visit www.ninjaresearcher.com/6051 to learn how to take your research skills and book report writing to the next level!

RESEARCH

DIGITAL LITERACY TOOLS

SEARCH LIKE A PRO
Learn about how to use search engines to find useful websites.

FACT OR FAKE?
Discover how you can tell a trusted website from an untrustworthy resource.

TEXT DETECTIVE
Explore how to zero in on the information you need most.

SHOW YOUR WORK
Research responsibly—learn how to cite sources.

WRITE

GET TO THE POINT
Learn how to express your main ideas.

PLAN OF ATTACK
Learn prewriting exercises and create an outline.

DOWNLOADABLE REPORT FORMS

Further Resources

BOOKS

Cockerham, Paul. *Porsche: The Ultimate Speed Machine*. Broomall, Pennsylvania: Mason Crest, 2018.

Garstecki, Julia. *Porsche 911 GT3*. Mankato, Minnesota: Black Rabbit Books, 2020.

Mason, Paul. *German Supercars: Porsche, Audi, Mercedes*. New York: PowerKids Press, 2019.

WEBSITES

Factsurfer.com gives you a safe, fun way to find more information.

1. Go to www.factsurfer.com.
2. Enter "Aston Martin Valkyrie" into the search box and click 🔍
3. Select your book cover to see a list of related websites.

Glossary

aerodynamics: the way air moves around objects. An aerodynamic design makes a car move faster.

carbon fiber: a very strong lightweight synthetic material. The body of a Valkyrie is made of carbon fiber.

cockpit: a space in a car where the driver sits to steer. The cockpit of the Valkyrie is teardrop shaped.

convertible: a car that has a roof that can be removed. The Valkyrie performance Spider is a convertible.

downforce: a combination of air resistance and gravity that acts on a moving vehicle, having the effect of pressing the vehicle down toward the ground. The Valkyrie is safer to drive because of the stability created by downforce.

hardtop: a car top that is hard and is not removable. The Valkyrie Coupe is a hardtop.

headquarters: the center or main part of a company. Aston Martin's headquarters is in Gaydon Village of Warwickshire, England.

horsepower: a unit of power that measures the power of an engine. A Valkyrie has a 1,000 horsepower engine.

V-12: a V-12 engine has 12 cylinders in the shape of a V. The Valkyrie has a V-12 engine.

Venturi tunnel: a Venturi tunnel optimizes airflow by directing it through the body shell of the car. The Valkyrie has a Venturi tunnel that goes under the floor of the cockpit.

Index

Aston Martin, 4, 6, 10, 12, 14, 16, 24
brake, 18
Coupe, 4, 9, 24
downforce, 8
engine, 6, 7, 9, 20
Formula One, 6, 18, 25

horsepower, 7
hybrid(s), 10, 20, 26
race car, 4, 12, 18
Spider, 4, 9, 24, 26
supercar(s), 4, 24, 25

PHOTO CREDITS

The images in this book are reproduced through: Atiketta Sangasaeng/Shutterstock 21 (bottom); all other images courtesy of Aston Martin Media (Max Earey 21)
Cover: Courtesy of Aston Martin Media, LiY/Shutterstock (background)

About the Author

James Savino lives in Florida with his family and his two crazy dogs. He's always loved anything with wheels that goes fast and has a loud motor. When he's driving down the road, he often rolls down his window just to hear the hum, roar, or growl of a good motor that pulls up beside him.